MYSTE

OF THE ANCIENT

THE LAST NEANDERTHALS

JOHN F. HOFFECKER

WEIDENFELD & NICOLSON
LONDON

The Neanderthals are one of the great mysteries of the human past. After almost a century and a half of discovery, analysis and debate, they remain an enigma. They are very close to us in many respects, but yet very distant. The Neanderthals are the last of our fossil ancestors to walk the earth, and are probably not true ancestors but actually evolutionary cousins who shared the planet with modern humans for more than 100,000 years. They seem to have been an alternative form of humankind – a variant of ourselves who possessed most of our traits but differed from us in some significant ways. Less than 30,000 years ago they vanished rather suddenly, and under suspicious circumstances.

The Neanderthals have become the most famous and most misunderstood of all fossil humans.

Neanderthal Origins

In order to understand how the Neanderthals fit into the story of human evolution, we must travel back to between six and ten million years ago. It is at this time that the earliest forms of humans (australopithecines) emerged in Africa, apparently in response to the shrinkage of tropical forests and corresponding growth of open landscapes. The earliest humans developed a trait that set them apart not only from their ancestral apes, but from most other mammals: the ability to walk upright on two legs. It is still not clear what advantage(s) upright locomotion conferred on the australopithecines, but by freeing the forelimbs for the making and using of tools, it set the stage for the next critical evolutionary step.

Slightly more than two million years ago the earliest members of the genus *Homo* appeared; they were the first humans to exhibit a pronounced increase in

brain size and the first to manufacture stone tools. Once again, the causes of these developments remain unclear, but their consequences were dramatic; within a relatively brief period of time, the toolmakers had expanded out of Africa across the southern half of Eurasia. Early *Homo* remains from Indonesia and southern China have recently been dated to between 1.7 and 2.0 million years ago.

By one million years ago populations of *Homo* had begun to move northward into cooler landscapes, as revealed by recent discoveries from Spain and the southern slope of the Caucasus Mountains. These were the first early humans to invade regions as far north as the city of Madrid (40° North); by half a million years ago, they had reached central Europe (50° North). As in the case of the expansion across southern Eurasia, the invasion of northern landscapes is associated with evolutionary change. During this period we see

Evidence for fire at Zhoukoudian. Reconstruction of a living floor based on information from the excavations at Zhoukoudian.

By half a million years ago early humans had moved northward into cooler lands. The remains of hearths found on this buried living floor in a north China cave reveal their mastery of fire.

The African australopithecines were the first human ancestors to walk upright, which freed their forelimbs for making and using tools and weapons.

the further increases in brain size that mark the gradual emergence of our own species *Homo sapiens*. It is from these archaic forms of *Homo sapiens* that both the Neanderthals and ourselves (*Homo sapiens sapiens*) evolved.

The first human populations to spread into northern Eurasia faced several new challenges to survival that seem to have been overcome through various behavioural and technological means (which may be related to the increases in brain size). Unfortunately, human skeletal remains and archaeological sites from this time range are rare, and there is limited information available for reconstructing their way of life.

The first remains of a Neanderthal to be recognized as an early form of human were discovered in 1856 in a cave in the Neander Valley in Germany.

The rockshelter of Le Moustier in southwestern France, which was discovered in 1863, was inhabited by Neanderthals during the Ice Age.

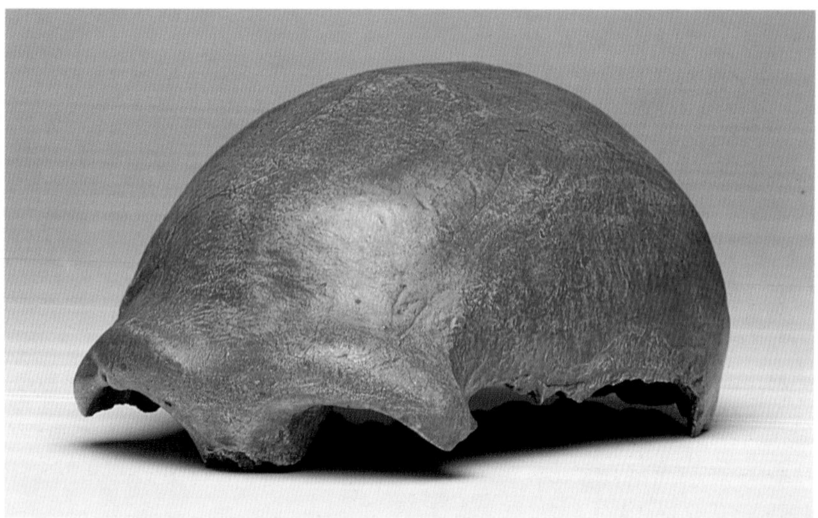

Control of fire was probably essential for coping with the colder temperatures of northern latitudes, and the oldest convincing traces of hearths have been found at sites in north China and central Europe dated to 500,000–350,000 years ago. The reduced abundance of edible plant foods must have required a greater reliance on animal foods, although microscopic analysis of teeth in a 500,000-year-old jaw from Germany revealed the severe wear of heavy plant food consumption. Evidence of big game hunting is scarce; most meat may have been scavenged from carcasses scattered across the landscape.

Within this broader picture of northern adaptation the Neanderthals emerged as a distinct human form roughly 200,000 years ago in Europe. As in the case of the appearance of the earliest *Homo sapiens*, the transition was a gradual one, and many of the characteristic Neanderthal traits evolved over a

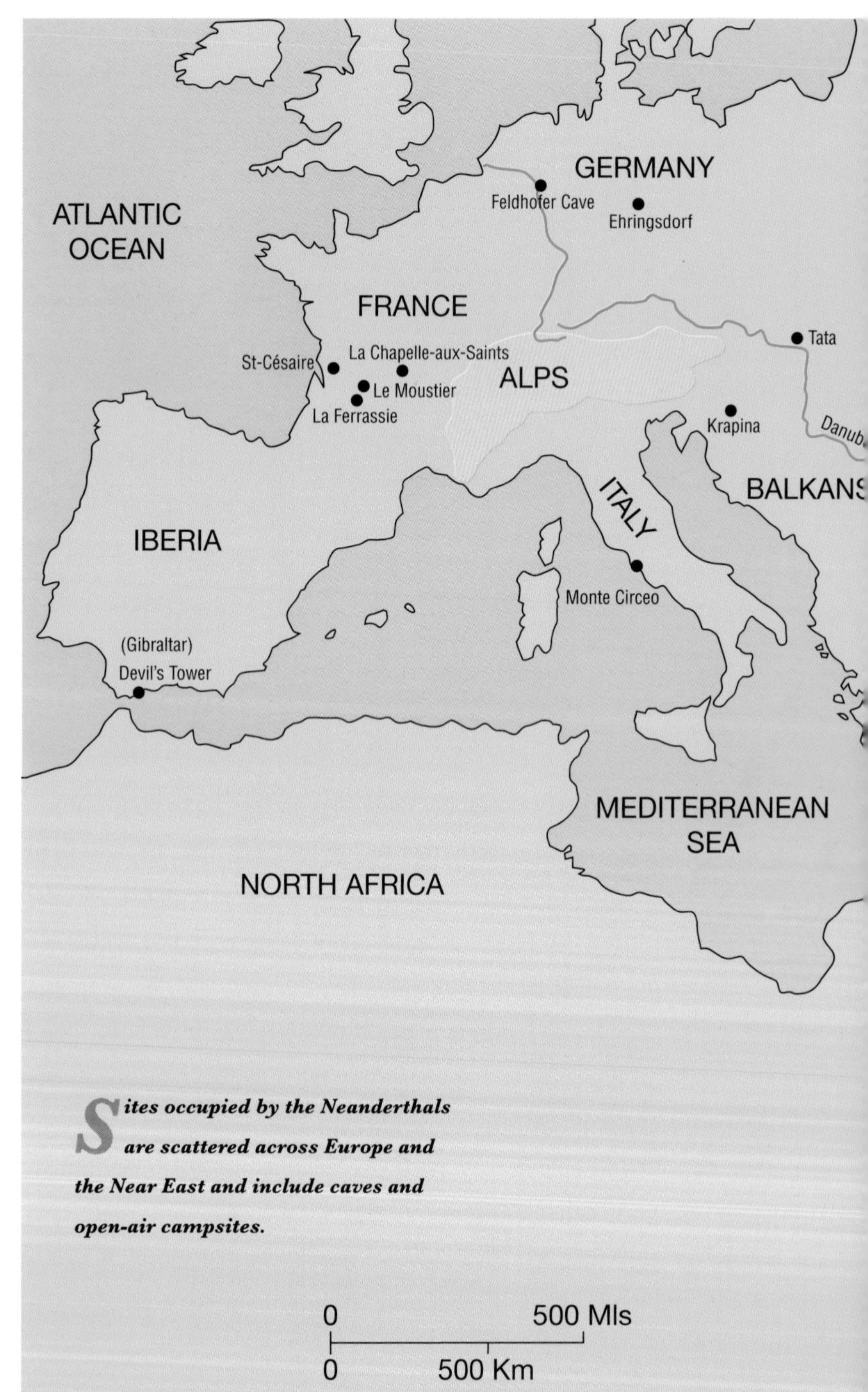

Sites occupied by the Neanderthals are scattered across Europe and the Near East and include caves and open-air campsites.

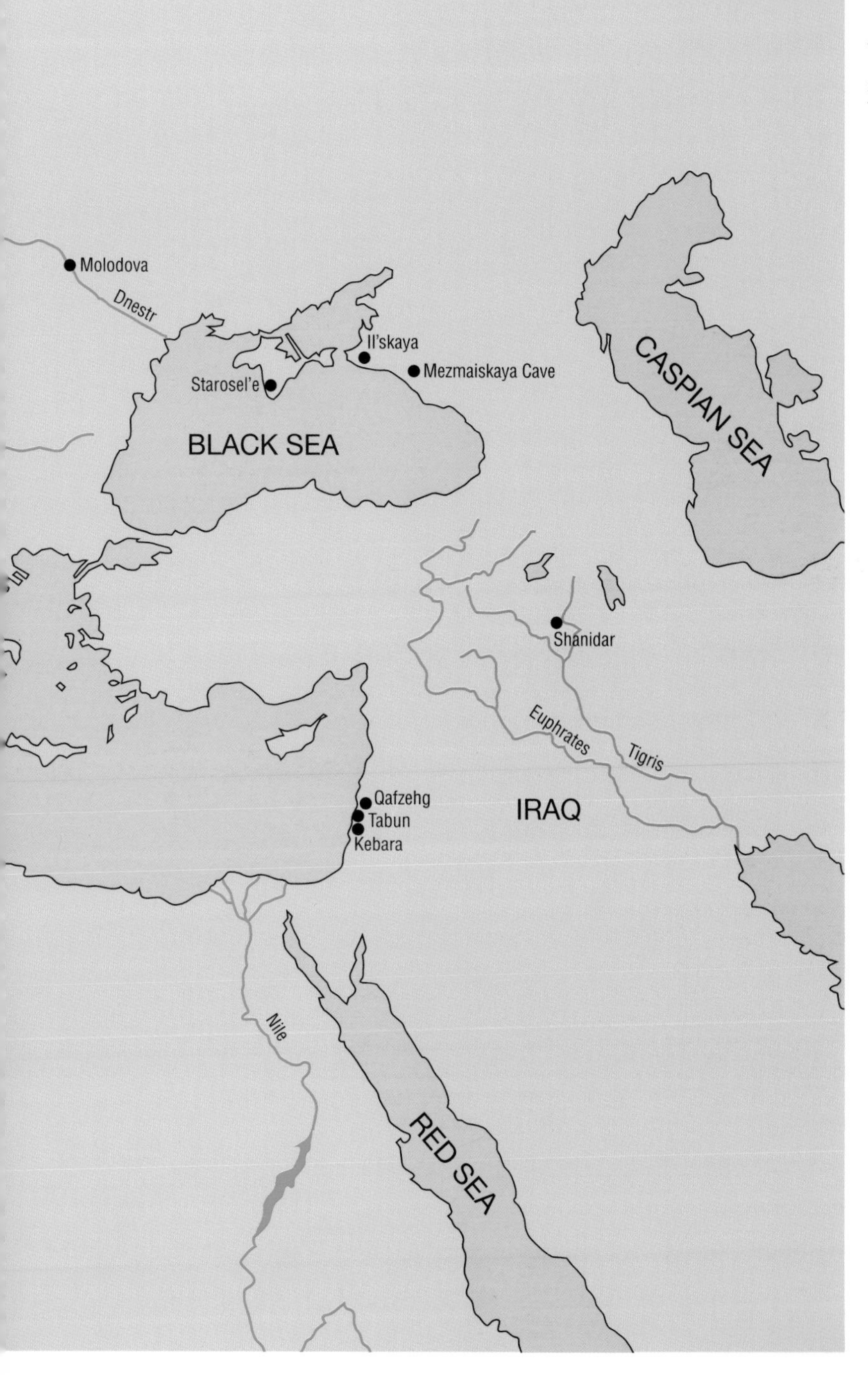
Molodova
Dnestr
Il'skaya
Mezmaiskaya Cave
Starosel'e
BLACK SEA
CASPIAN SEA
Shanidar
Euphrates
Tigris
Qafzehg
Tabun
Kebara
IRAQ
Nile
RED SEA

long period of time. The Neanderthals (who are classified scientifically as *Homo sapiens neanderthalensis*) represented a significant advance over their predecessors in their ability to cope with northern environments. This is strikingly evident in the distribution of their remains in space and time. The Neanderthals were the

Neanderthals occupied this campsite beside a stream in southern Russia.

first humans to occupy Europe during periods of intense glacial cold; earlier human populations seemed to have abandoned higher latitudes during glacial periods. They were also the first

מוסטרית

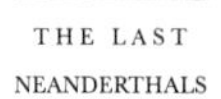

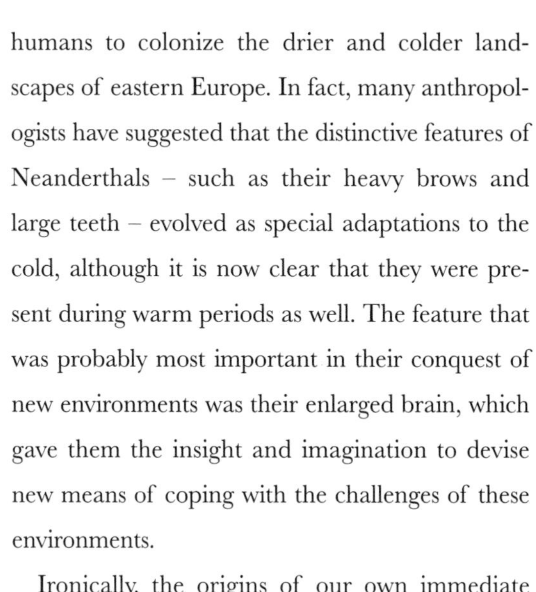

humans to colonize the drier and colder landscapes of eastern Europe. In fact, many anthropologists have suggested that the distinctive features of Neanderthals – such as their heavy brows and large teeth – evolved as special adaptations to the cold, although it is now clear that they were present during warm periods as well. The feature that was probably most important in their conquest of new environments was their enlarged brain, which gave them the insight and imagination to devise new means of coping with the challenges of these environments.

Ironically, the origins of our own immediate ancestors (early *Homo sapiens sapiens*) remain less clear that those of our Neanderthal cousins. The lineages appear to have split at some point before 200,000 years ago, and the two forms of humankind shared the planet until roughly 30,000 years ago, when the Neanderthals vanished. While the Neanderthals occupied Europe throughout this period, our own ancestors resided in Africa. The Near East became a 'crossroads' inhabited at various times by both groups, although it is not certain that the two populations ever actually co-existed together for any length of time.

At Tabun Cave in Israel, Neanderthal remains have been dated to roughly 100,000 years ago or earlier.

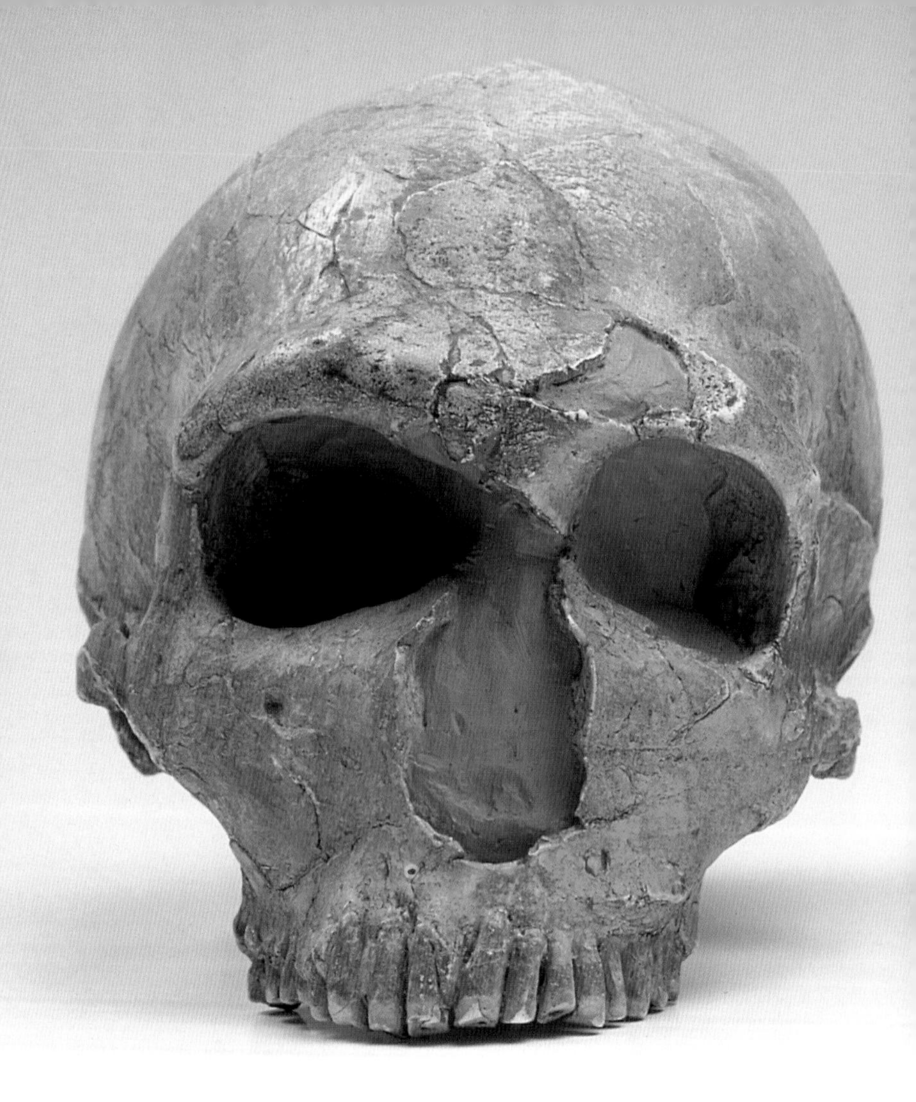

Anatomy of a Neanderthal: Brains and Muscle

Unlike earlier human forms, the Neanderthals are represented by a relative abundance of skeletal remains. This is chiefly a function of their comparatively recent age; erosion and weathering destroy most remains and the sites that contain them – including caves – in a few hundred thousand years. Anthropologists have thus acquired a wealth of bones and teeth, which they have sub-

A comparison of the skull of a Neanderthal (right) with a modern human (left) reveals the peculiar features of the former.

jected to exhaustive studies. As a result, we know a great deal about the anatomy and appearance of our late cousins, who possessed a remarkable combination of brains and muscle.

We know that the Neanderthals were highly robust and muscular, a trait that they shared with their more primitive predecessors. They were not especially tall

The Neanderthals possessed a striking combination of powerful muscles and a large brain.

(adult males seem to have averaged about 1.60 m in height), but they possessed large joints, thick leg-bones, and many parts of the skeleton exhibit the deep marks of powerful muscle attachments. The bones of their children show that the robust skeleton began to develop at an early age, apparently to endure a life of heavy physical demands; most adult skeletons also reveal signs of disease and injury.

The head that rested on this muscular frame was even more striking in contrast to ourselves, and unique among human forms. The heavy brow-ridges and flattened top were also traits inherited from their predecessors, and ones that have contributed significantly to their primitive and brutish image. But the Neanderthals combined these traits with an exceptionally large brain that exceeds even our own in total volume. While modern human brains average roughly 1,400 cubic centimetres, our sample of fossil Neanderthal skulls indicates an average brain volume of over 1,500 cubic centimetres! The elongated shape of these skulls has encouraged past speculation that the Neanderthal brain was organized differently from our own, and without comparable development of areas of higher thought. Today most anthropologists recognize that there is an insufficient basis for such conclusions, and that we must turn to other sources of information for an understanding of the Neanderthal mind.

The Neanderthal face also reflected some unique features that must have appeared odd, perhaps repulsive, to the modern humans who eventually met up with them. It possessed inflated cheeks and a remarkably prominent nose, but lacked a chin. The front teeth were excessively large and, along with the cheek teeth, were placed so far forward relative to the rest of the face that a

This skeleton from Kebara Cave in Israel revealed new information about the anatomy of the Neanderthals.

gap existed between the last molar and the back of the jaw. Studies of the teeth reveal extreme wear, including microscopic scratches that seem to have been caused by the regular practice of cutting of materials such as animal hide held firmly in the grip of the jaw.

The Economic and Social Order

Evidence that Neanderthal daily life may have been more stressful and dangerous than that of their modern human contemporaries leads us to wonder how much it differed from the latter. Did the Neanderthals rely more heavily on brute strength and endurance than on careful planning and technology? Anthropologists are sharply divided over this issue, either believing that the Neanderthal economy was fundamentally different from that of the modern human populations who succeeded them, or that it was quite similar.

This cave in the northern Caucasus Mountains of Russia was probably used by Neanderthal groups during warmer months as a base while hunting sheep and bison.

The bones of large animals found in Neanderthal caves and camp-sites sometimes exhibit cuts made by stone tools.

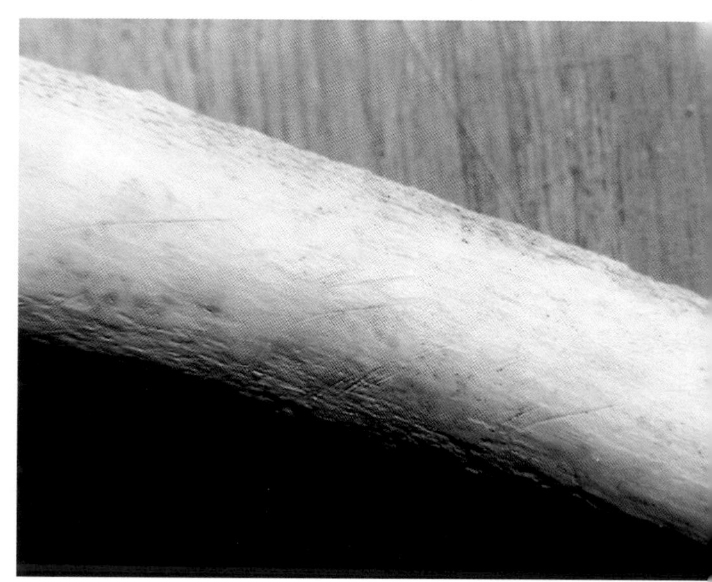

Like all peoples of the Ice Age (and some of the post-Ice Age), the Neanderthals did not practise agriculture, but pursued an economy based on nomadic foraging. Some anthropologists have suggested that the Neanderthals lacked the advance planning and scheduling of modern human foraging people, and that they wandered from place to place, often depending on chance encounters with animal prey and other sources of food. However, the study of artefacts and animal bones from the sites that they occupied reveals a pattern fundamentally similar to that of modern foragers. For example, on the

The Neanderthals inhabited an Ice-Age world populated by mammoths, cave bears and other large animals that have since vanished.

slopes of the northern Caucasus Mountains in Russia, we find a diverse array of sites at varying elevations used for different purposes during different seasons of the year. The Neanderthals seemed to have visited these locations at specific times to procure locally and seasonally available foods, which must have required planned and scheduled movements and activities. There is reason to believe, however, that they moved around within smaller territories than their modern successors.

Some anthropologists have also suggested that the Neanderthals were less competent and effective hunters than modern humans. The settlement of northern environments poor in edible plant foods would have demanded a heavy reliance on animal foods, but these might have been obtained from the scavenging of carcasses rather than hunting live prey. There are sites in central Italy that appear to reflect at least partial reliance on scavenging deer carcasses. On the other hand, sites in France, Russia and elsewhere contain strong evidence for the hunting of medium and large mammals such as red deer and bison, the bones of which exhibit the marks of butchering by stone tools and lack the characteristics of the remains of animals that have succumbed to disease or non-human predators. Some Neanderthal sites even contain large quantities of the remains of herd mammals that might have been driven into ravines or over cliffs in the manner of a Indian bison hunt on the plains of North America.

The question of Neanderthal economy is inextricably linked to that of their society. The highly flexible social organization of modern human foragers – as revealed by studies of hunting and gathering peoples like the south African

Careful excavation and study of Neanderthal camps and caves has revealed much information about the diet and economy of their occupants.

Bushmen and the Eskimo – allows response to variations in the timing and location of available food sources. Thus, a small group of adult males might be despatched to the mountains for a protracted winter hunt, while a large group of males, females and children might be assembled temporarily on the plains to provide the necessary numbers for a bison herd drive in the late summer. Did Neanderthal society possess this organizational flexibility?

Once again, some anthropologists have argued that there were radical differences between the Neanderthals and modern humans. It has long been suggested that the Neanderthals lacked an incest taboo – universal among all modern human societies as far as we know – and did not follow an established pattern of mating and marriage between nuclear families (exogamy). Exogamy ensures a network of co-operative alliances among family groups that provides the foundation for modern human society, and seems to be an important part of the organizational flexibility of modern human foraging peoples. But anthropologists have found it difficult, if not impossible, to reconstruct Neanderthal society from the limited variety of remains that are available for study. To shed any light on this issue, we must turn to another aspect of

Neanderthal social organization remains a mystery. It may have been very different from that of modern human hunting and gathering peoples.

Neanderthal life, and one that reveals the most dramatic and important contrast between that life and our own.

Language and Culture?

The critical factor in the shaping of our economic and social order is language and the use of symbols. The intricate network of alliances among families and the complex economic strategies of modern human foragers rest on language and culture – on our ability to formulate and communicate abstract concepts and shared beliefs through symbols. The development of the capacity for using symbols was the last and most significant event in the evolution of modern humans, and in the millennia following the Ice Age, as many populations began to settle down to an agricultural way of life, it provided the basis for civilization.

Despite their large brains and impressive foraging skills, there is much evidence that the Neanderthals led an existence that was largely if not wholly devoid of symbols. While the modern humans who succeeded them in Europe 30,000 years ago left behind a rich legacy of ornament and art (including spec-

***T**he walls of European caves occupied by Neanderthals lack the colourful paintings created by their modern human successors.*

***T**he Neanderthals manufactured and resharpened an array of simple stone tools.*

tacular paintings on the walls of caves), the Neanderthals left us virtually nothing in this respect. There are isolated examples of objects bearing simple designs, including two engraved bone fragments from caves in Bulgaria and France. Another French cave yielded a drilled fox tooth that might have been an ornament, and the possible fragment of a flute has been reported recently from Slovenia.

Anthropologists have searched for regional differences among the stone tools produced by Neanderthals that might reflect the sort of cultural variations often evident in even the simplest of modern human artefacts. They have found remarkably little variation among tool types from different parts of Europe and the Near East. Instead, some have suggested that differences in the percentages of these generic tool types might reflect different Neanderthal cultural traditions, but recent studies of Neanderthal tools indicate that much of this variation is likely due to the degree to which an individual tool has been worn and resharpened.

There is evidence that the Neanderthals buried their dead in caves, which would seem to indicate some element of ritual and belief in their lives, but the interpretation of these burials remains controversial. Some of the skeletons seem to represent individuals who were simply buried in falling rubble and debris; whole or partial skeletons of cave-dwelling bears and other animals are not uncommon. The careful excavation of a Neanderthal skeleton discovered in France several years ago revealed no evidence of burial pit. Some Neanderthal remains have been found associated with stone tools and other objects thought to have been grave goods; one burial found in Iraq even yielded traces of flowers (in the

The Neanderthals were probably the first humans to bury their dead, and numerous examples of apparent graves have been found in Europe and the Near East.

form of pollen concentrations) that may have been tenderly placed next to the deceased. Many anthropologists suspect that the association of these objects with the skeletons is largely fortuitous. Once again, the contrast with their modern human successors – who sometimes buried their dead with pendants, necklaces, and sculptures – is a stark one.

There have also been attempts to deduce the linguistic abilities of the Neanderthals directly from the reconstructed anatomy of their vocal tracts. These studies reveal some differences with modern humans that suggest that the Neanderthals may not have been able to produce a comparable range of speech sounds. Because no modern human language employs the full range of sounds that we are capable of making, the implications of these studies are not clear, but they do seem consistent with the general lack of evidence for the use of symbols.

Kebara Cave in Israel was probably visited by Neanderthals about 60,000 years ago.

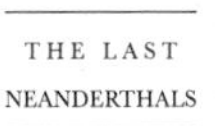

The End of the Neanderthals

The last known Neanderthal to walk the earth died in southern Spain roughly 30,000 years ago, and the location of this sombre event may be significant. From that day onwards only the modern apes – safely hidden in the tropical forests of Africa and Asia – remained as living reminders of the last ten million years of our evolution. The fate of the Neanderthals is still hotly debated among anthropologists worldwide. Were they destroyed by modern human populations, or did they mingle with the newcomers and contribute their heritage to the living peoples of western Eurasia?

Genetic studies indicate that living human populations of Europe and Asia are derived from an ancestral African population. Modern humans were present at least 100,000 years ago in Africa, and appeared shortly thereafter in the adjoining Near East. Many anthropologists believe that these people spread across Europe and Asia, replacing the Neanderthals and other archaic forms of *Homo sapiens* roughly 50,000–30,000 years ago. In fact, a widely publicized

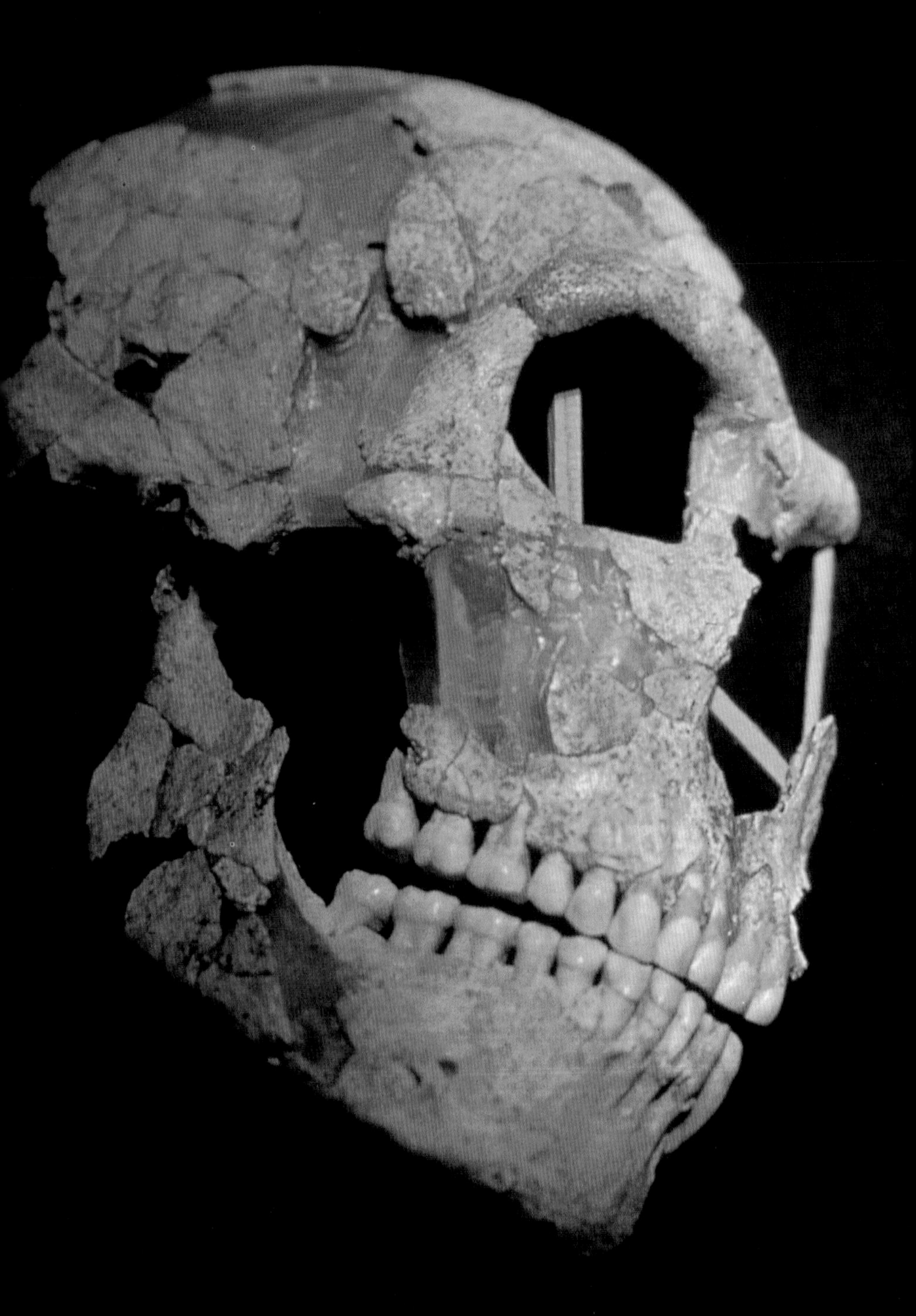

genetics study of 1987 suggested that all living humans are descended from a single African woman (the 'Eve Hypothesis'). Modern humans appear in eastern Europe over 40,000 years ago, and subsequently seem to move westward towards the Atlantic coast. Although some anthropologists insist that the modern humans of Europe are descended – at least to some extent – from the Neanderthals, there is little evidence of intermediate or hybrid forms (especially in western Europe), and the genetic contribution of the latter to the former would seem to be minimal at best.

If it came down to a physical contest between the two, it is hard to imagine the powerful Neanderthals being overwhelmed by their rather effete modern human counterparts. However, the invaders may have triumphed over the former through superior organization and technology. Both the physical appearance and accompanying evidence for art and symbols indicate that the people who entered Europe roughly 40,000 years ago were fully modern in their behaviour. Our studies of the Neanderthal economy show that the two populations must have competed for the same resources.

A curious twist to this story is that, before their arrival in Europe, modern humans revealed no signs of this new behavioural pattern; their sites are as devoid of art and symbols as those of the Neanderthals. The change in behaviour seems to have been related to the move north. The catalyst may have been the onset of the last major cold period of the Ice Age, approximately 60,000 years ago. For all their skills in coping with northern environments, the Neanderthals seem to have had problems adjusting to the especially harsh conditions in eastern Europe. Recently dated fossils document a Neanderthal intrusion

The youngest Neanderthal skeletons in western Europe have been found with simple ornaments and stone tools similar to those made by modern humans.

into the Near East at this time, which some believe to have been made by refugees from glacial Europe. If so, the Neanderthals may have won the first round, temporarily displacing the modern human inhabitants of the region.

During the next 20,000 years modern humans moved into eastern Europe, simultaneously manifesting their new use of symbols. It may be that the latter conferred advantages in communication and organization that were essential to settlement of glacial landscapes where resources had become scarce and widely dispersed; there is evidence (in the form of materials moved from one place to another) that the modern humans were foraging across much larger territories than the Neanderthals. Eventually, groups of modern humans probably intruded on lands still occupied by the latter. The encounter between the two groups, which has been the subject of various novels and films, seems to have been protracted and complex. Some late Neanderthal sites in western Europe contain simple ornaments and

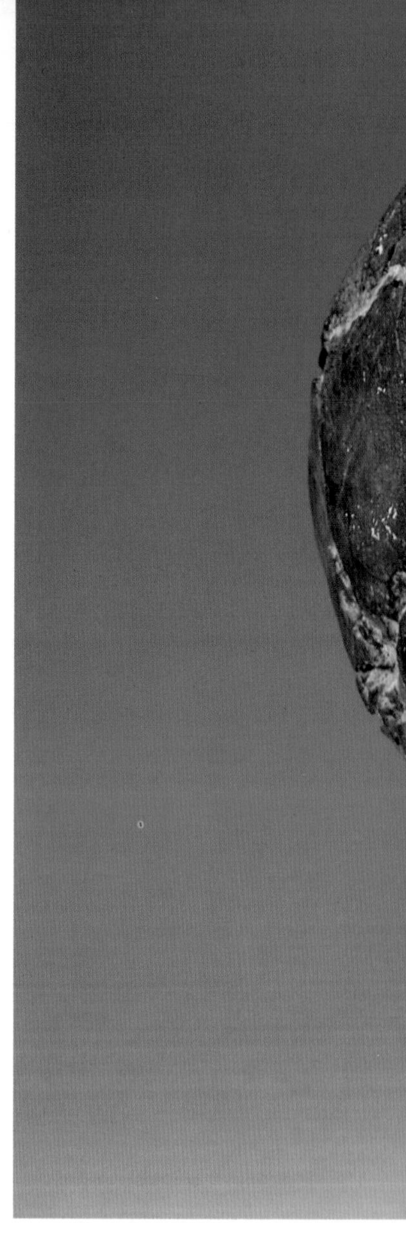

Anatomically modern humans were living in the Near East roughly 100,000 years ago.

tools similar to those of modern humans, suggesting that in their final period they were beginning to adopt some of the ways of the invaders. There are no traces of violence on the Neanderthal skeletons found to date, and the cause of their disappearance remains a mystery.

PHOTOGRAPHIC ACKNOWLEDGEMENTS
Cover E.T. Archive; pp. 3, 4–5, 6-7, 8, Natural History Museum, London [NHM]; pp. 8–9 Paul G. Bahn; p. 12–13 John Hoffecker [JH]; pp. 14–1 5 Jane Callander [JC]; pp. 16–17 NHM; p. 18 ETA; pp. 19, 20–21 NHM; pp. 22–3, 23 JH; pp. 24–5 AKG London; p. 26 JC; pp. 28–9 NHM; p. 30 Paul G. Bahn; p. 31t,b JH; pp. 32–3 ETA; pp.34–5 JC; p.36 Paul G Bahn; pp. 38–9 ETA.

First published in Great Britain 1997
by George Weidenfeld and Nicolson Ltd
The Orion Publishing Group
5 Upper St Martin's Lane
London WC2H 9EA

A CIP catalogue record for this book is available from the British Library
ISBN 0 297 823108

Picture Research: Joanne King

Designed by Harry Green

Typeset in Baskerville